Pocket Safety Guide
Confined Spaces

First edition published 2010

© *Witherby Seamanship International Ltd, 2010*

British Library Cataloguing in Publication Data
A catalogue record for this book is available from the British Library.

ISBN 13: 978 1 905331 81 9

Printed and bound in Great Britain by Bell & Bain Ltd, Glasgow

Published in 2010 by:

Witherby Seamanship International Ltd
4 Dunlop Square, Livingston
Edinburgh, EH54 8SB
Scotland, UK

Tel No: +44(0)1506 463 227
Fax No: +44(0)1506 468 999

Email: info@emailws.com
www.witherbyseamanship.com

Contents

1 Types of Space 1

  1.1 What is a Confined Space? 1

  1.2 Permit Entry Confined Space 2

2 Reasons for Entering a Confined Space 3

  2.1 Is it Necessary? 3

  2.2 Reasons for Entering a Confined Space 3

  2.3 Recognition 3

3 Hazards 4

  3.1 Physical Hazards 5

  3.2 Control 6

4 Hazardous Atmospheres and Methods of Testing for Them 7

  4.1 Hazardous Atmospheres 7

    4.1.1 Flammable atmospheres 7

    4.1.2 Toxic atmospheres 8

    4.1.3 Irritant atmospheres 8

    4.1.4 Asphyxiating atmospheres 9

  4.2 Testing Hazardous Atmospheres 10

    4.2.1 Testing flammable atmospheres 11

    4.2.2 Testing toxic atmospheres 12

5 Ventilation, Gas Freeing 14

6 Entry Checks 15

7   Duties   17

   7.1   Training for Attendant   17

   7.2   Duties of the Person Authorising or in
       Charge of the Entry   17

8   Communication   19

   8.1   Requirements of a Confined Space Entry
       Communication System   20

9   Equipment   22

10 Rescue   23

11 Case Studies   24

   11.1 Case Study #1   24

   11.2 Case Study #2   24

   11.3 Famous Last Words?   24

12 Precautions   27

   12.1 Testing, Evaluation and Monitoring   27

1   Types of Space

There are many types of confined spaces, but no matter what the type they all have something in common. They have limited ways to get in and out, and the atmosphere within them could be dangerous.

1.1   What is a Confined Space?

Generally speaking, a confined space is an enclosed or partially enclosed space that:

- Is not primarily designed or intended for human occupancy
- has a restricted entrance or exit because of the location, size or method
- can represent a risk for the health and safety of anyone who enters, because of one or more of the following factors:
  - the design, construction, location or atmosphere
  - the materials or substances in it
  - work activities being carried out in it
  - mechanical, process and safety hazards present.

A confined space, despite its name, is not necessarily small.

1.2   Permit Entry Confined Space

Permit entry confined space means an enclosed space that:

- Is large enough and laid out in such a way that a worker could enter and perform work
- has limited means of entry and exit such as a storage bin, hopper, vault, pit or diked area
- is not designed for continuous occupancy by the worker
- has one or more of the following characteristics:
  - contains or may contain a hazardous atmosphere
  - contains the potential for engulfment by loose particles
  - has an internal layout such that someone entering could be trapped or asphyxiated by inwardly converging walls or a floor that slopes downward and tapers to a smaller cross-section
  - contains any other recognised serious safety or health hazard.

2 Reasons for Entering a Confined Space

2.1 Is it Necessary?

Is it absolutely necessary that the work is carried out inside the confined space?

In many cases where there have been deaths in confined spaces, the work could have been carried out elsewhere.

2.2 Reasons for Entering a Confined Space

Entry is usually necessary to perform a function such as inspection, repair, maintenance (cleaning or painting) or similar operations. It would normally be an infrequent or irregular function.

2.3 Recognition

Seafarer training is essential if there is to be recognition of what constitutes a confined space and the hazards that may be encountered in it. This training should stress that death to a seafarer is a likely outcome if proper precautions are not taken before entry is made.

3   Hazards

All hazards found in a regular workspace can also be found in a confined space, where they are likely to become even more hazardous.

Hazards in a confined space can include:

- Poor air quality: there may be an insufficient amount of oxygen for a worker to breathe. The atmosphere might contain a poisonous substance that could make the worker ill or even cause the worker to lose consciousness. Natural ventilation alone will often not be sufficient to maintain breathable quality air
- chemical exposures due to skin contact or ingestion as well as inhalation of 'bad' air
- fire hazards: there may be an explosive/flammable atmosphere due to flammable liquids and gases and combustible dusts which if ignited lead to fire or explosion
- process-related hazards such as residual chemicals or the release of contents of a supply line
- noise
- safety hazards such as moving parts of equipment, structural hazards, slips or falls
- radiation
- temperature extremes, including atmospheric and surface
- shifting or collapse of bulk material
- barrier failure resulting in a flood or release of a free-flowing solid
- uncontrolled energy, including electrical shock
- poor visibility
- biological hazards
- trapping/pinch points.

3.1   Physical Hazards

The statistics for confined space accidents indicate that accident rates in confined space working and rescue are relatively high compared with other industrial activities. Fatality figures indicate that 1-2 rescuers die for each worker's life lost in a confined space accident.

The confined space may present the following risks to both workers and would-be rescuers:

- Being overcome by gas, fumes vapour or lack of oxygen
- injury due to fire or explosion
- being drowned or buried under free-flowing solids
- being overcome due to high temperatures.

When your body temperature exceeds 39°C, you will become less efficient and will be prone to heat exhaustion, heat cramps or heat stroke.

From the review of data from a number of case studies, it can be concluded that fatalities occurred as a result of encountering one or more of the following potential hazards:

- Lack of natural ventilation
- oxygen deficient atmosphere
- flammable/explosive atmosphere
- unexpected release of hazardous energy
- limited entry and exit
- dangerous concentrations or air contaminants
- physical barriers or limitations on movement
- instability of stored product.

In each of these cases, there was a lack of **Recognition** and **Testing, Evaluation,** and **Monitoring** prior to entry, nor had a well-planned **Rescue** been attempted.

3.2   Control

One of the most difficult aspects to control is that of unauthorised entry, particularly where there are large numbers of workers and trades involved, such as welders, painters, electricians and safety monitors. This is a situation frequently found at dry-dock.

4 Hazardous Atmospheres and Methods of Testing for Them

4.1 Hazardous Atmospheres

Hazardous atmospheres encountered in confined spaces can de divided into four distinct categories:

- Flammable
- toxic
- irritant/corrosive
- asphyxiating.

4.1.1 Flammable atmospheres

A flammable atmosphere generally arises from enriched oxygen atmospheres, vaporisation of flammable liquids, by-products of work, chemical reactions, concentrations of combustible dusts and desorption of chemicals from the inner surfaces of the confined space.

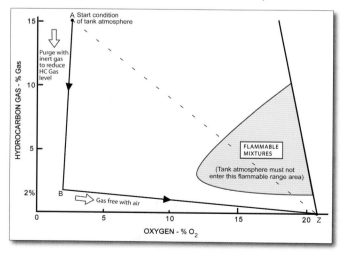

Flammable Range Diagram

An atmosphere becomes flammable when the ratio of oxygen to combustible material (hydrocarbons) in the air is neither too rich nor too lean for combustion to occur. Combustible gases or vapours will accumulate when there is inadequate ventilation in areas such as a confined space.

The by-products of work procedures can generate flammable or explosive conditions within a confined space. Specific kinds of work, such as spray painting, can result in the release of explosive gases or vapours. Welding in a confined space is a major cause of explosions in areas that contain combustible gas.

4.1.2 Toxic atmospheres

The substances to be regarded as toxic in a confined space can cover the entire spectrum of gases, vapours and finely divided airborne dust. The sources of toxic atmospheres encountered may arise from the following:

- The manufacturing process (for example, in producing polyvinyl chloride (PVC), hydrogen chloride is used as well as vinyl chloride monomer, which is carcinogenic)
- the product stored (removing decomposed organic material from a tank can liberate toxic substances, such as H_2S)
- an operation performed in the confined space (for example, welding or brazing with metals capable or producing toxic fumes)
- during loading, unloading, formulation and production, mechanical and/or human error may also produce toxic gases that are not part of the planned operation.

4.1.3 Irritant atmospheres

Irritant or corrosive atmospheres can be divided into primary and secondary groups. Primary irritants affect the surface of the body tissue, whereas a secondary irritant is one that may produce systemic toxic effects (ie effects on the entire body) in addition to the surface irritation. Examples of primary and secondary irritants are shown in the table below.

Primary irritants	Secondary irritants
Chlorine	Benzene
Ozone	Carbon tetrachloride
Hydrochloric acid	Ethyl chloride
Sulphuric acid	Trichloroethane
Nitrogen dioxide	Trichloroethylene
Ammonia	Chloropropene
Sulphur dioxide	

Irritant gases vary widely among all areas of industrial activity. They can be found in plastic plants, chemical plants, the petroleum industry, tanneries, refrigeration industries, paint manufacturing and mining operations.

Prolonged exposure at irritant or corrosive concentrations in a confined space may produce little or no evidence of irritation and may cause a general weakening of the defence reflexes from changes in sensitivity. The danger in this situation is that the worker is usually not aware of any increase in their exposure to toxic substances.

4.1.4 Asphyxiating atmospheres

Reduction of oxygen in a confined space may be the result of either consumption or displacement.

Normal atmosphere is composed of approximately 20.9% oxygen, 78.1% nitrogen and 1% argon, with small amounts of various other gases. Reduction of oxygen in a confined space may be the result of either consumption or displacement.

Decreased oxygen levels (below the atmospheric level of 20.9% by volume) can cause various effects including:

Level of 17%:   increased breathing volume and accelerated heartbeat

Between 14-16%:   increased breathing volume, accelerated heartbeat, very poor muscular coordination, rapid fatigue and intermittent respiration

Between 6-10%:   nausea, vomiting, inability to perform and unconsciousness

Less than 6%:   spasmatic breathing, convulsive movements and death within minutes.

The consumption of oxygen takes place during combustion of flammable substances, as happens during welding, heating, cutting and brazing. Oxygen may also be consumed during chemical reactions, such as during the formation of rust on the exposed surfaces of the confined space (iron oxide).

Another cause of oxygen deficiency is its displacement by another gas. Examples of gases that are used to displace air, and therefore reduce the oxygen level, are helium, argon and nitrogen. The use of nitrogen to inert a confined space has claimed more lives than carbon dioxide.

4.2   Testing Hazardous Atmospheres

It is important to understand that some gases or vapours are heavier than air and will settle to the bottom of a confined space and that some gases are lighter than air and so will be found at the top. Therefore, it is necessary to test all areas (top, middle and bottom) of a

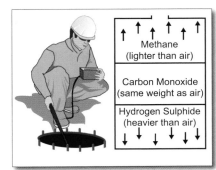

confined space with properly calibrated testing instruments to determine the gases that are present.

Before entry, it is necessary to test the atmosphere in the confined space for oxygen levels, flammability and/or any contaminants that may be present. This testing must be carried out by a qualified person using equipment that has been approved for

use in such areas. The testing equipment itself should be checked to make sure it is working properly before use. The manufacturer's recommended procedures must be followed.

If test results conclude that the atmospheric condition of the confined space is unacceptable, entry is prohibited until conditions are brought within acceptable limits. This may be done by purging, cleaning and/or ventilating the space. Purging is the method by which gases, vapours or other airborne impurities are displaced from a confined space. The confined space may also

be made non-flammable, non-explosive or otherwise chemically non-reactive by displacing or diluting the original atmosphere with steam or gas that is non-reactive within that space, a process referred to as 'inerting'.

4.2.1 Testing flammable atmospheres

Flammable gases, such as acetylene, butane, propane, hydrogen, methane, natural or manufactured gases, or vapours from liquid hydrocarbons, can be trapped in confined spaces and, since many gases are heavier than air, they will settle in lower levels as in pits, sewers and various types of storage tanks and vessels.

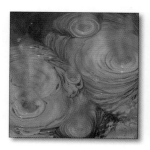

In a closed top tank, lighter than air gases may rise and develop a flammable concentration if trapped above the opening.

4.2.2 Testing toxic atmospheres

Carbon monoxide (CO) is a hazardous gas that may build up in a confined space. This odourless, colourless gas has approximately the same density as air and is formed from incomplete combustion of organic materials such as wood, coal, gas, oil and gasoline. It can also be formed from microbial decomposition of organic matter in sewers, silos and fermentation tanks. Carbon monoxide is an insidious toxic gas because of its poor warning properties. Early stages of CO intoxication are nausea and headache. Carbon monoxide may be fatal at 1000 ppm in air, and is considered dangerous at 200 ppm because it forms carboxyhaemogloben in the blood, which prevents the distribution of oxygen in the body.

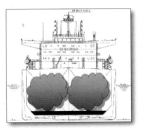

CO is a relatively abundant colourless, odourless gas, so its presence in any untested atmosphere should be suspected. It must also be noted that a safe reading on a combustible gas indicator does not ensure that CO is not present, and it must be tested for specifically. The formation of CO may result from chemical reactions or work activities, so fatalities due to CO poisoning are not confined to any particular industry. There have been fatal accidents in sewage treatment plants due to decomposition products and lack of ventilation in confined spaces.

The air within the confined space should be tested *from outside of the confined space* before entry into the confined space. Care should be taken to ensure that air is tested throughout the confined space, side to side and top to bottom. A trained worker using detection equipment that has remote probes and sampling lines should carry out the air quality testing. The sampling should show that:

- The oxygen content is within safe limits, not too little and not too much
- a hazardous atmosphere (toxic gases, flammable atmosphere) is not present
- ventilation equipment is operating properly.

Atmospheric testing is required for two distinct purposes:

- Evaluation of the hazards of the permit space
- verification that acceptable entry conditions exist for entry into that space.

5   Ventilation, Gas Freeing

Ventilation by a blower or fan may be necessary to remove harmful gases and vapours from a confined space.

It should always be considered that the most unfavourable situation exists in every confined space and that the danger of explosion, poisoning and asphyxiation will be present at the onset of entry.

Before forced ventilation in initiated, information such as restricted areas within the confined space, voids, the nature of the contaminants present, the size of the space, the type of work to be performed and the number of people involved should be considered.

The ventilation air should not create an additional hazard through recirculation of contaminants, improper arrangement of the inlet duct or by the substitution of anything other than fresh (normal) air (approximately 20.9% oxygen, 78.1% nitrogen and 1% argon with small amounts of various other gases).

6   Entry Checks

- ☐ Is it necessary?
- ☐ Are the instruments used in atmospheric testing properly calibrated?
- ☐ Was the atmosphere in the confined space tested?
- ☐ Was oxygen at least 20.8% - not more than 21%?
- ☐ Were toxic, flammable or oxygen-displacing gases/vapours present?
  - ➤ Hydrogen sulphide   ppm
  - ➤ Carbon monoxide   ppm
  - ➤ Methane   LEL
  - ➤ Carbon dioxide   ppm
- ☐ Will the atmosphere in the space be monitored while work is going on?
- ☐ Has the space been ventilated before entry?
- ☐ Will ventilation be continued during entry?
- ☐ Is the air intake for the ventilation system located in an area that is free of combustible dusts and vapours and toxic substances?
- ☐ Has the space been isolated from other systems?
- ☐ Has electrical equipment been locked out?
- ☐ Have lines under pressure been blanked and bled?
- ☐ Is special clothing required (boots, chemical suits, glasses, etc)?
- ☐ Is special equipment required (eg rescue equipment, communications equipment etc)?
- ☐ Are special tools required (eg spark proof)?
- ☐ Is respiratory protection required (eg air-purifying, supplied air, self-contained breathing apparatus, etc)?

☐ Will there be a standby person on the outside in constant visual or auditory communication with the person inside?

☐ Will the standby person be able to see and/or hear the person inside at all times?

☐ Has the standby person(s) been trained in rescue procedures?

☐ Will safety lines and harness be required to remove a person?

☐ Are you familiar with emergency rescue procedures?

☐ Do you know who and how to notify in the event of an emergency?

☐ Has a confined space entry permit been issued?

7   Duties

7.1   Training for Attendant

Any worker functioning as an attendant at a
permit entry confined space must be trained in
the emergency action plan, the duties of the
attendant and in:

- Proper use of the communications
  equipment used for communicating
  with authorised workers entering the confined space or for
  summoning emergency or rescue services
- authorised procedures for summoning rescue or other emergency
  services
- recognition of the unusual actions of a worker which could
  indicate that they might be experiencing a toxic reaction to
  contaminants which could be present in that space
- any training for rescuers, if the attendant will function as a
  rescuer
- any training for workers who enter the confined space, if the
  permit specifies that the duty of the attendant will rotate among
  the workers authorised to enter the confined space.

7.2   Duties of the Person Authorising
or in Charge of the Entry

The person who authorises or is in charge of
the permit entry confined space must comply
with the following:

- Make certain that all pre-entry
  requirements, as outlined on the permit,
  have been completed before any worker
  is allowed to enter the confined space

- make certain that any required pre-entry conditions are present
- if an in-plant rescue team is to be used in the event of an emergency, make sure that they would be available
- make sure that any communication equipment that would be used to summon either the in-plant rescue team or other emergency assistance is operating correctly
- terminate the entry upon becoming aware of a condition or set of conditions whose hazard potential exceeds the limits authorised by the entry permit.

8 Communication

Communication between the worker inside and the standby person outside is of utmost importance. If the worker should suddenly feel distressed and not be able to summon help, an injury could quickly become a fatality. Frequently, the body positions that are assumed in a confined space make it difficult for the standby person to detect an unconscious worker.

When visual monitoring of the worker is not possible because of the design of the confined space or location of the entry hatch, a voice or alarm-activated explosion-proof type of communication system will be necessary.

Suitable illumination of an approved type is required to provide sufficient visibility for work.

Noise in a confined space, which may not be intense enough to cause hearing damage, may still disrupt verbal communication with the emergency standby person on the exterior of the confined space.

An adequate communication system will be needed and should enable communication:

- Between those inside the confined space
- between those inside the confined space and those outside
- to summon help in case of an emergency.

8.1 Requirements of a Confined Space Entry Communication System

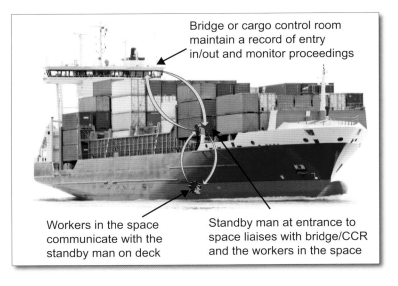

Bridge or cargo control room maintain a record of entry in/out and monitor proceedings

Workers in the space communicate with the standby man on deck

Standby man at entrance to space liaises with bridge/CCR and the workers in the space

- Reliable
- two-way continuous speech communication capability
- compact, rugged, environmentally protected from water ingress
- rapid deployment capability with simple, intuitive operation and minimum training requirements
- IS certification covering all anticipated gases, vapours and dusts
- shift plus battery life
- compatible with hard hats, gloves and breathing apparatus
- operation possible with hazardous material (Hazmat) isolation suits
- any cable reels taken into the entry to be compact and lightweight
- roving capability for attendant to permit local mobility while monitoring transmissions
- emergency alarm button on the entrants' equipment
- hands-off voice operated operation (where appropriate)

- ability to operate in a high noise environment, with either headset, earpiece or telephone handset, as necessary.

Transmission discipline must be effective. The aim is to communicate any message in a logical format with good enunciation and then to smoothly hand over the communications channel. This is essential in 'press to talk' systems. Message confirmation should also be practised to confirm that instructions or information have been fully understood.

9   Equipment

There will be situations where full breathing apparatus must be worn. This apparatus comprises:

Self-contained breathing apparatus (SCBA)

This is generally open circuit apparatus with a cylinder of compressed air supplying air to the user via a full face mask. Various sizes are available, providing working times of up to 45 minutes. These times are shorter where there is high heat stress. In fully saturated air temperatures of 36°C for example, safe wearing time may be no more than 20 minutes. Under these circumstances, the tasks must be accomplished quickly and effectively.

Airline breathing apparatus

Used where working times beyond 45 minutes are required. Airline breathing apparatus are fresh air, constant flow and demand flow systems, sometimes used in conjunction with backup cylinders. Demand flow airline systems can probably at best provide 100 metres range, with simple fresh airlines limited to less than 10 metres. A further alternative is closed circuit SCBA, offering 2 hours or more duration. However, this requires specialist breathing apparatus training.

10 Rescue

A standby person should be assigned to remain on the outside of the confined space and be in constant contact (visual or oral) with the workers inside. The standby person should not have any other duties except to serve as a standby, with knowledge of who should be notified in case of an emergency. Standby personnel should not enter a confined space until help arrives, and then only with proper protective equipment, lifelines and respirators.

> *Over 50% of the workers who die in confined spaces are attempting to rescue other workers.*

Rescuers must be trained in and follow established emergency procedures including use of equipment and techniques (lifelines, respiratory protection, standby persons, etc). Steps for safe rescue should be included in all confined space entry procedures. Rescue should be well-planned and drills should be frequently conducted on emergency procedures. Unplanned rescue, such as when someone instinctively rushes in to help a downed co-worker, can easily result in a double fatality, or even multiple fatalities if there is more than one would-be rescuer.

Since deaths in confined spaces often occur because the atmosphere is oxygen deficient or toxic, confined spaces should be tested prior to entry and then monitored continually.

Entry with hoist and standby personnel

11   Case Studies

11.1 Case Study #1

On 4th October 1984, two workers (26 and 27 years old) were overcome by gas vapours and drowned after rescuing a third worker from a fracturing tank at a natural gas well. The tank contained a mixture of mud, water and natural gas. The first worker had been attempting to move a hose from the tank to another tank. A chain secured the hose and when the worker moved it, the chain fell into the tank. The worker entered the tank to retrieve the hose and was overcome.

11.2 Case Study #2

On 13th May 1985, a 21-year-old worker died inside a wastewater holding tank, that was 4 feet in diameter and 8 feet deep, while attempting to clean and repair a drain line. Sulphuric acid was used to unclog a floor drain leading into the holding tank. The worker collapsed and fell face down into 6 inches of water at the bottom of the tank. A second 21-year-old worker attempted a rescue and was also overcome and collapsed. The first worker was pronounced dead at the scene and the second worker died 2 weeks later. Cause of death was attributed to asphyxiation by methane gas. Sulphuric acid vapours may have also contributed to the cause of death.

11.3 Famous Last Words?

"I just calibrated our gas monitor last month - it should still be accurate"

Calibration is the cornerstone of any successful gas-monitoring programme. Unless you compare and adjust your instrument to a known concentration of gas prior to each use, you have no assurance of the unit's accuracy. OSHA[1] points out in *29 CFR 1910.146 (page 4551, section (ii)(c))*

---

[1] United States Department of Labor, Occupational Safety and Health Administration

that workers should be using a calibrated, direct reading gas monitor. Unfortunately, they make no mention of calibration frequency. Refer to the manufacturer's recommendations, and set up a programme according to their guidelines. For your protection, and the protection of your employer, always document calibrations and preserve these records in a log. The calibration log should be maintained, current and include all the gas monitors used. These records should be on file for no less than 1 year, although retention for 5 years is advised.

"I put the tubing into the space, turned on the stamping pump, waited the recommended time (for the unit to respond) and all my readings checked out OK"

Monitoring the gases in one area of a confined space is a dangerous oversight. Some gases are heavier than air, some are lighter. The molecular weight of a gas determines where it will accumulate within a confined space. Air has a molecular weight of approximately 28.8. Common gases, such as hydrogen sulphide and methane, will stratify naturally at different levels. Hydrogen sulphide, for example, has a molecular weight of 34 so it can be expected to accumulate closer to the floor. Methane will be present closer to the top of the space due to its molecular weight of 16. This is assuming, of course, the space has poor natural ventilation or has not been stirred up and the gases allowed to stratify. *29 CFR 1910.146 (page 4557, Appendix B, section 4)* confirms this by mandating that employees monitor the gas in 4-foot intervals in every direction of travel. For vertical entries, this is easy, simply use flexible tubing and start at the top and work your way down. For horizontal samples, a rigid probe at least 4 feet in length will be needed to safely address the situation.

"Our inerting operation looks like a success - our instrument shows 1.2% oxygen and no LEL gases"

This common assumption can prove fatal. If you are using most common gas monitors to evaluate the concentration of combustible gases (LEL) in an inert environment, you may be getting a false sense of security. Gas monitors employing catalytic diffusion sensors are very effective in determining hazard potential because they operate on the fire triangle principle. When you remove one of the elements from the fire triangle, namely oxygen, no reaction takes place. The danger in this is if you have explosive amounts of gas in an inert situation and if oxygen is introduced or if the gas were to leak out of that area, a hazard could result.

"The manual says that Teflon tubing should be used for sampling, but it's so hard to work with. It really doesn't matter because our confined spaces are safe. We have been using this black neoprene hose for years and never detect any hazardous gas"

Some reactive gases, such as chlorine, nitrogen dioxide, ammonia, styrene, etc, can be absorbed into the walls of sampling tubing. The longer the tubing, the more gas can be 'scrubbed' from the sample. If this absorption takes place, little or no gas will get to the instrument's sensors and the hazard will go undetected. Teflon tubing has very smooth walls that gases cannot adhere to. Advances in Teflon tubing technology have yielded several flexible designs that are resistant to kinks so the days of Teflon tubing being hard to work with are long gone.

"When I attach the sampling pump to our gas monitor and turn on the pump, as long as I can hear the motor running I am feeding the instrument a good sample"

With the flow monitoring technology today, it is easier than ever to ensure proper sampling of confined spaces. Sampling pumps with built in flow monitors are precise and easy to use. Care should be taken to use only sampling pumps that monitor gas flow through the pump. Most sampling pumps only monitor the vacuum pressure at the inlet of the pump, so pump assembly failures can go undetected. This is a serious safety concern. Along with flow monitoring, audible and visual failure warnings are ideal for enhanced hazard recognition.

12 Precautions

These are the things you should be aware of before you enter a confined space:

- Know how to enter it safely
- know how to exit quickly
- know that the atmosphere in the space is tested and found to be free of dangerous levels of toxic or flammable vapours, and that there is sufficient oxygen
- know that the atmosphere within the space is going to remain safe while you are working
- know the rescue plan in the event of an emergency, and make sure the proper rescue equipment is available and in good condition
- know that another person outside the confined space is keeping an eye on you as you work, and that they also know the rescue plan
- know what other procedures are necessary to follow to work safely, such as locking out energy sources.

To completely isolate a confined space, the closing of valves is not sufficient. All pipes must be physically disconnected or isolation blanks bolted in place. Other special precautions must be taken in cases where flammable liquids or vapours may re-contaminate the confined space.

12.1 Testing, Evaluation and Monitoring

All confined spaces should be **Tested** by a qualified person before entry to determine whether the confined space atmosphere is safe for entry. Tests should be made for oxygen level, flammability and known or suspected toxic substances. **Evaluation** of the confined space should consider the following:

- Methods for isolating the space by mechanical or electrical means (ie double block and bleed, lockout, etc)

- lockout/tagout procedures
- ventilation of the space
- cleaning and/or purging
- work procedures, including use of safety lines attached to the person working in the confined space and its use by a standby person if trouble develops
- personal protective equipment required (clothing, respirator, etc)
- special tools required
- communications system to be used.

The confined space should be continuously **Monitored** to determine whether the sphere has changed due to the work being performed.